AF588072

Great Hornbill

by Grace Hansen

Abdo Kids Jumbo is an Imprint of Abdo Kids
abdobooks.com

abdobooks.com

Published by Abdo Kids, a division of ABDO, P.O. Box 398166, Minneapolis, Minnesota 55439.

Abdo Kids Jumbo™ is a trademark and logo of Abdo Kids.

Printed in the United States of America, North Mankato, Minnesota.

102020

012021

Photo Credits: iStock, Science Source, Shutterstock

Production Contributors: Teddy Borth, Jennie Forsberg, Grace Hansen
Design Contributors: Dorothy Toth, Pakou Moua

Library of Congress Control Number: 2020910698

Publisher's Cataloging-in-Publication Data

Names: Hansen, Grace, author.

Title: Great hornbill / by Grace Hansen

Description: Minneapolis, Minnesota : Abdo Kids, 2021 | Series: Asian animals | Includes online resources and index.

Identifiers: ISBN 9781098205942 (lib. bdg.) | ISBN 9781098206505 (ebook) | ISBN 9781098206789 (Read-to-Me ebook)

Subjects: LCSH: Birds--Juvenile literature. | Exotic birds--Juvenile literature. | Rain forest animals--Juvenile literature. | Animals--Juvenile literature. | Asia--Juvenile literature. | Endangered species--Juvenile literature.

Classification: DDC 598.2913--dc23

Table of Contents

Great Hornbill Habitat

Great hornbills are found in Nepal, India, and a few other countries. They mainly live in tall, wet, evergreen forests.

Asia
Nepal
India
Malaysia
Indonesia

Body

Great hornbills are large birds. They can grow up to 50 inches (127 cm) long. They are the heaviest hornbill **species**.

A great hornbill's **wingspan** can be 60 inches (152 cm) long! The flapping of its wings can be heard from far away.

Great hornbills have black, white, and yellow feathers. They have a solid black band on their white tails.

The bird got its name for the great **casque** that sits on its bill. A female's casque and bill are almost entirely yellow. A male's casque and bill are orange, white, black, and yellow.

Food

Fruit is the great hornbill's favorite food. The bird also eats small mammals, reptiles, and insects.

Baby Great Hornbills

Great hornbills are **social**.

They can be found in pairs.

They also live in small family groups and smaller flocks.

Before a female lays her eggs, she finds a **hollow** tree. She builds a nest in its trunk. Then she seals herself inside using mud and dung.

The male feeds the female through the hole. She does not come out of the tree until her chicks have feathers.

More Facts

- Great hornbills lay 1 to 2 eggs at a time.
- The giant **casque** helps a great hornbill make louder noises.
- The bird's casque looks heavy, but it is actually very light.

Glossary

casque – on hornbills, the horny helmet-like growth that sits above the beak.

hollow – having an empty space on the inside.

social – living in groups instead of alone.

species – a group of living things that look much alike and can have young with one another but not with those of other groups.

wingspan – the distance from the tip of one wing to the tip of the other.

Index

Visit **abdokids.com** to access crafts, games, videos, and more!

Use Abdo Kids code

AGK5942

or scan this QR code!